Enjoy!

Griffith Moon

Printed in the United States of America
First Printing, 2017

Published by Griffith Moon
Santa Monica, California
GriffithMoon.com

ISBN: 978-0-9981686-4-7

www.tiffanyshlain.com

*To my mother Carole Lewis & my father Leonard Shlain,
who always said raising children was like
growing their own best friends.*

*To my husband Ken,
who makes me feel like anything is always possible.*

*To our children, Odessa & Blooma,
who taught me how to be a mother.*

"...the best invention in the ... is the mind of a child."

– Thomas Edison

CONTENTS

INTRODUCTION :
THE FUTURE STARTS HERE

I've always believed the technology we create is a direct extension of us, not something separate. As Marshall McLuhan wrote in *The Medium is the Message*, first published in 1967, "The wheel is an extension of the foot, the book is an extension of the eye, clothing, an extension of the skin, electric circuitry, an extension of the central nervous system."

We're constantly striving to extend our reach. When we couldn't see far enough, we made the telescope. When we needed to speak with people who weren't within shouting distance, we created the telephone. When we wanted to travel to far-off places in a short period of time, we invented the airplane. When we needed to connect data, we developed the Internet. Since its arrival, the Internet has evolved to do so much more than just connecting data: It now links ideas, knowledge, feelings, emotions, and memories among people all over the world. The Internet, is, in many ways, an extension of our brains — an extension of us. And, like us, these inventions can be both good and bad, and everything in between.

Over two billion people from around the world regularly engage with this global Internet brain, allowing innumerable and diverse perspectives to interact. The creative potential of all these intersecting ideas is extraordinary. Yet the

Internet is still relatively young as a generally accessible form of human interaction. It's still in a phase of rapid growth and change. Compared with the human life cycle, the Internet is still in its formative years. This means that, as with a developing child's brain, we must pay careful attention to how we are developing our global brain.

But how far can that parallel go? What is the connection between the developing brain of a human child and the emerging global brain of the Internet? Can the lessons learned from observing the former affect how we nurture the growth of the latter? What can we do, every day, to strengthen both? What small, regular activities promote our children's healthy brain development? What daily habits make for smart, mindful Web usage that will lead to mindful participation, as well as to a stronger foundation for the growth of the Internet?

The idea to write a book that connected these two lines of inquiry arose when I was traveling for my documentary *Connected*, a film that explores what it means to be connected — emotionally, technologically, and socially — in the 21st century. After screenings in theaters and at conferences, schools, and community centers around the world, one question came up over and over: What is all this technology doing to our brains?

Around this same time, one of my mentors began sharing recent research on the development of children's brains that uses a new technology, magnetoencephalography, or MEG imaging. With MEG imaging, we can actually see what's happening inside a child's mind during its first five years, a glimpse that confirms how crucial these years are to human brain development. This research – and the questions asked by *Connected* audiences – inspired me and my team to make a new film, *Brain Power*, that investigates these new methods for nurturing a young child's mind — and how they might help us improve the way we use technology and develop the Internet.

This book is an outgrowth, an extension of that short film. It explores more deeply how our collective brains are coevolving with technology. It roughly follows the script of *Brain Power*, the film, with video excerpts, and links to research, talks, and other films that further unpack its ideas. The format is unique — which means it's an experiment — and lets us explore topics more comprehensively than a conventional film allows for. To make the experience more interesting, I recommend watching the 10-minute film first at www.letitripple.org.

We began *Brain Power* by delving into brain-development research, and we quickly discovered that, quite fittingly, the terms neuroscientists use to describe the growth of a child's brain (connections, links, overstimulation) are the same ones we reach for when describing aspects of the ever-evolving Internet.

We also learned that our brains change throughout life because behavior, experiences, and environment can alter neural pathways and synapses. This phenomenon, called "neuroplasticity," means that old dogs can, in fact, learn new tricks. But the young can learn a lot more. A baby's brain is extremely neuroplastic, developing rapidly from every interaction it has.

At its MEG Brain Imaging Center, the University of Washington's Institute for Learning and Brain Sciences (I-LABS) operates one of the most advanced MEG-scanning centers in the world, under the direction of Patricia Kuhl, the center's founder and co-director. MEG scanners, which are safe and noninvasive, let us watch the brains of infants and children while they're listening to their mothers or looking at an object or person. MEGs have shown us that even day-old newborns are already transmitting powerful neural signals and establishing neural pathways.

A baby's brain has an enormous number of neurons – a hundred billion, the same number an adult brain has — but most of the connections between all those neurons aren't yet there. Those connections happen as a baby interacts with the surrounding world. As MEG imaging shows, connections are happening from the first hour of an infant's life. By the age of 3, a young child's brain has made so many connections, it has three times more than an adult does. Experiences strengthen connections, causing some to remain while others are pruned. Our role as nurturers in this critical stage of development — the first 2,000 days, or from birth to 5 years old — is to provide an environment that strengthens as many connections as possible and prunes the ones that aren't needed, ultimately building a strong foundation for the rest of the child's life.

As neuroscience reminds us, the beginning of any project is a crucial time. It's when the foundation for future success is laid. If we're at the metaphoric first 2,000 days of life for the Internet, then right now is when we need to pay close and careful attention to developing its brain. When the moment comes that every person on the planet who wants to be online is able to be (I believe that the majority of you will live to experience this), we will surely hope that we had spent the time to develop the Internet in the best way possible.

As a society, we've been doing a lot of hand-wringing about how distracted we've become, how skimming over and snacking on ideas has replaced deep thinking. But it's important to remember that this same kind of concern cropped up many times before in the past, during transformational shifts in the way we communicate. When the written word was invented, Socrates worried it would make us lose our memory. When I was growing up, many believed that television was going to turn our brains to "mush." In every new medium, something is lost and something is gained. My father, Leonard Shlain, was a surgeon who wrote best-selling books about the relationships between literacy and patriarchy, between art and science, and trends in history — all with a special focus on the brain. He often cited a quote from Sophocles: "Nothing vast enters the lives of mortals without a curse." [1]

Brain Power combines four of my greatest passions: family, filmmaking, understanding human behavior, and the evolution of the Internet. My husband, Ken Goldberg, and I have two children. Blooma, 3, is at the most rapid stage of brain development. Odessa, 9, is honing her higher-order thinking skills, such as planning and flexibility. My mother, Carole Lewis, a psychologist, has instilled in me a fascination with emotional development and has taught me much about how thinking around human interaction has evolved. Discussions over the years with both my parents about how our brains grow have helped shape both my brain and my interest in the brain. Now, as a mother, it's amazing to see the process in action. You're watching a human happen, and it's miraculous.

The Web holds equal interest for me. For a long time now, it's been a huge part of my life. In 1996, I had the opportunity to create the Webby Awards to recognize the best of the Internet. I spent nearly a decade tracking the evolution of the Web and honoring its innovators. I have also spent more than two decades making films, many of which explore the coevolution of technology and society. My films are accompanied by in-depth "discussion kits," featuring books, conversation cards, websites, mobile apps, and any other available tools for stretching ideas and triggering conversations. TED Books is part of this quest to use the most innovative resources out there for exploring ideas and telling stories in new ways.

I always describe the process of writing my films like this: the lines we end up with are distilled from a rich sauce of ideas that have been marinating, for many years, in lots of research and many discussions with my favorite thinkers. *Brain Power* is cowritten with my filmmaking collaborator at the Moxie Institute, Sawyer Steele, and with my collaborator in life, my husband, Ken, who's a new-media artist and robotics professor at the University of California–Berkeley. In many ways, this book embodies the key ingredients that Sawyer, Ken, and I put in the sauce. These chapters aim to help all of us think about the best ways to develop our children's brains, along with the global brain of the Internet.

Bon appetit!

Tiffany Shlain
Mill Valley, California, 2012

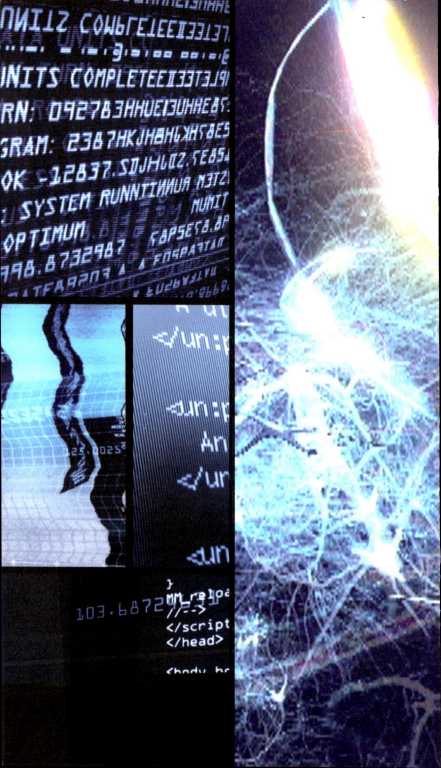

OUR BRAINS, REWIRED

I'm walking down the street on the cusp of spring. I pass a tree branch that's starting to bloom with hundreds of purple flowers. I hear laughter from a somewhat distant woman walking toward me, and I think of one of my favorite lines from a poem by Ralph Waldo Emerson: "The earth laughs in flowers." As I get close to her, I realize she's been laughing in response to a text, and that her head has been down the whole time. I feel two emotions simultaneously. I want to tell her, "Stop! Look at that beautiful branch of flowers!" But I also think, "How lovely that someone just made her laugh." She's connecting with someone somewhere else.

Wrestling with the good, the bad, and the potential of all this technology is my constant state of existence. The technology we've created – and that now dominates our work and home lives — allows us to experience something new. But it's also taking something away from us: being in the present moment.

It seems that many people are on one side of this argument or the other. Either technology is good and will solve everything, or technology is bad and will destroy civility and real connection. I want to explore the space in the middle that allows us to extol the mind-bending facts of what technology allows us to do while also acknowledging the moments we're losing, the flowers we're not seeing and smelling.

While technology has given us many opportunities we've never had before, it's also putting new kinds of stress in our lives. This brings us back to asking what all this technology is doing to our brains. Perhaps it would be better to ask how it's shaping our brains.

A hormone, dopamine, provides one of the clues. Neuroscientists have studied dopamine since 1958, when Arvid Carlsson and Nils-Ake Hillarp, working at the National Heart Institute of Sweden, discovered how the chemical functions in the brain. Dopamine plays an influential role in mood, attention, memory, understanding, learning, and reward-seeking behavior. That last function is critical. Dopamine is what makes us seek pleasure and knowledge. It's what makes us search, be it for food, sex, or just information. This leads to something known as a dopamine-induced loop. Have you ever been up late, unable to stop linking from website to website? Or maybe you compulsively text, tweet, or email. These are dopamine-induced loops. Dopamine makes you seek the initial connection. Then, when you're rewarded with a response or with more information, dopamine is what makes you want more. So you click again, and again, and again. [2] A recent *New York Times* article gave the German name for this phenomenon: sehnsucht, or "addictive yearning." [3]

In the absence of this dopamine surge, people start to feel bored and inattentive, which can have bad consequences: overlooking an email, for instance, or the interruption of creativity and deep thought — or, worse still, getting into a car accident. Researchers at the National

Institute on Drug Abuse have compared the lure of digital stimulation to that of drugs and alcohol. [4] A study at the University of California–Irvine found that people interrupted by email had higher levels of stress than those who weren't interrupted and were able to focus. Dr. Gary Small, a professor in the University of California–Los Angeles's psychiatry department, tells us that stress hormones have been shown to reduce short-term memory. [5]

That's some of the bad part, but there's also a positive side. We have never been more stimulated by new ideas, more able to connect with people we love and with those who share our interests. When I'm working on a script and am simultaneously online, following the links people throw out on Twitter or YouTube or Facebook, new ideas mix in with whatever I was just writing. When I refocus from such distractions and return to my writing, I suddenly have fresh perspectives, fresh infusions of ideas, to bring to my work. In an essay on Edge.org, Paul Kedrosky says that using the Internet is like:

> having a private particle accelerator on my desktop, a way of throwing things into violent juxtaposition, and then the resulting collisions reordering my thinking. The result is new particles — ideas! — some of which are BDTs (big, deep thoughts), and many of which are nonsense. But the democratization of connections, collisions, and therefore thinking is historically unprecedented. We are the first generation to have the information equivalent of the Large Hadron Collider for ideas. [6]

The Internet is truly an unprecedented tool for discovery and new ideas.

Technology also gives us tools for telling our own stories in new ways. Before the Internet and social networks, media tended to be unidirectional — a few people controlled a few stories to be consumed by many. Today, networks allow us to share through infinite channels. But this, too, has a hormonal effect. Paul Zak, a neuroeconomist at Claremont Graduate University, has found that social networking produces a burst of oxytocin, the hormone responsible for bonding, empathy, trust, and generosity. I sometimes imagine that the Internet is flooding the planet with oxytocin. Every text, tweet, and email produces the hormone that makes us more empathetic and more inclined to share and collaborate. Maybe all this oxytocin is why collaboration is on the rise?

Technology is rewiring the human brain. It's happened before. Five hundred years ago, the latest and greatest technology was the printing press. With that startling invention came increased and widespread literacy. In his book *The Alphabet Versus the Goddess*, my father argued that literacy rewired the brain, too. [7] It strengthened the linear, logical, "masculine" left hemisphere, he said, and weakened the more visual, fluid, big-picture, "feminine" right hemisphere. He believed that this shift, privileging the male over the female, led to a cultural shift as well, resulting, ultimately, in patriarchy and misogyny. Then, in the 19th century, there was another shift. He called it the "iconic revolution": the invention of photography and electromagnetism, which led to film, television, and computers. My father hoped this more visual technology would rewire the brain yet again — only, this time, privileging the right hemisphere over the left, the feminine over the masculine, which would move us back to a cultural equilibrium. These were the stories I grew up on.

Three hundred years ago, books printed on Gutenberg's press inspired controversy and debate. Would the spread of books also spread sedition and discontent? But culture marched on, and society didn't crumble. It won't crumble now. Mathew Ingram, a blogger for Gigaom.com, writes:

Is there a need for moderation when it comes to phones or the Internet or social media? Of course there is, and social norms are developing around those things, just as they developed around the horseless carriage and the telephone and plenty of other modern inventions. One of the devices that has historically drawn the most criticism from scholars and theologians for its corrupting effect on humanity seems to have worked out pretty well – it's called the book. If we can figure that out, I'm sure we can figure out how to handle cellphones and status updates.

TUNING IN AND OUT

So how will the Internet affect the human brain and human culture as a whole? That story is still being played out. In a 2009 study, published in the journal *Cell Death and Differentiation*, Italian scientists found that physical activity, social interaction, and multisensory stimulation all affect the central nervous system — both in terms of turning on certain genes and increasing the growth of cells, especially those associated with the visual system of the cerebral cortex. [8] For example, watching the film *Brain Power* (and reading this book) is reshaping the connections in your brain right now. But because we humans are the ones creating this technology, we are equally responsible for how we use it.

We can choose when we use technology, and we can choose when to turn it off. We can also choose when to focus our attention on the things that are deeply important to us: being truly present with the people we love.

In 2008, my father was diagnosed with brain cancer. During his illness, I began to think a lot about time — about how little of it we have, about how meaningless connections are unless we connect deeply. But connecting deeply requires attention and being present. I didn't want to be distracted when I was with my father, so I'd turn off my cellphone. Later, Ken and I decided to institute something we'd been attempting since we'd met: One day a week, we'd turn off the technology in our lives. We called it our "technology Shabbat." From sundown on Fridays to sundown on Saturdays, we shut down every cellphone, iPad, TV, and computer in the house. This practice has been profoundly life-changing for us. It resets my soul each week. Seriously. Inspired to not only unplug ourselves, but also to invite others to try it, Ken and I participated in an event called National Day of Unplugging, making a two-minute film for the occasion called *Yelp: With Apologies to Allen Ginsberg's "Howl."* You can watch it at www.moxieinstitute.org

In his book *The Sabbath*, published in 1951, the Jewish philosopher Abraham Heschel describes the period of rest and worship as "a cathedral in time," a concept that resonates when you unplug from technology. [9] During our technology Shabbats, time slows down. Albert Einstein said that "time is relative to your state of motion." With all this texting, tweeting, posting, and emailing, we're making our minds move faster, which accelerates our perception of time. It seems there isn't a day that goes by when I don't end up thinking, "How did it get to be 5 p.m.?"

When my family unplugs, time starts to move at this beautiful, preindustrial pace. And what is the day you want to feel extra-long? Saturday. Our Saturdays now feel like four days of slow living that we savor like fine wine. We garden, we ride our bikes, we cook, and I write in my journal. I actually read (one thing at a time). I can have a thought without being able to immediately act on it. I can think about someone without being able to contact them at that moment. I have found that it's good to let a thought sit. It changes when you don't act on it. For one day a week, I like letting my mind go into a completely different mode. On our Saturdays, we're able to enjoy all those activities that seem to get pushed aside by the allure of technology. Because we're neither Orthodox nor Amish, we do use a car; we do turn on the lights and answer our landline for emergencies. We've given a modern interpretation to a very old idea of the Sabbath, but we still try to be as unavailable as possible to the world — with the exceptions of each other and our children.

There's another benefit to this weekly unplugging. By sundown on Saturday, we can't wait to get back online: We're hungry for connection. We appreciate technology all over again. We marvel anew at our ability to put every thought and emotion into action by clicking, calling, and linking.

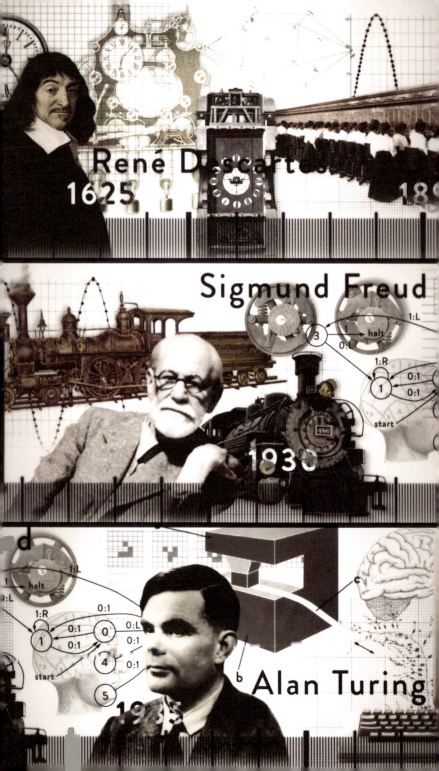

René Descartes
1625

18

Sigmund Freud
1930

Alan Turing
1935

HOW WE KNOW
WHAT WE KNOW

"The more I learn, the more I realize how little I know." Plato attributed this to Socrates. Lao Tse said, "The more you know, the less you understand." And the writer Will Durant stated, "Education is a progressive discovery of our own ignorance." [10] This is especially true when it comes to our collective understanding of the human brain.

Not long after we humans began to think, we started thinking of ways to describe our brain. Many thinkers throughout history have used the technologies of their day as metaphors aimed at understanding how the brain works. For the ancient Egyptians, the brain was like metalworking; they compared it to slag. For Galen, in the second century, the brain was a hydraulic system. [11] By the late 1620s, René Descartes, the French philosopher and mathematician best known for his statement "I think, therefore I am," viewed the brain as a clock. "Mind and body are alienated from each other like the mechanism of a clock and the hours it shows," he wrote in *Treatise of Man*. [12]

During the Industrial Revolution, the brain was described as a factory full of gears, levers, and pistons (presumably operated by Dickensian ragamuffins). And, for Sigmund Freud, the mind was a steam engine. "Human behavior is the result of opposing forces," he wrote, describing what he believed drove the human psyche. "Pressures build up and are released and 'expressed' causing us to eventually act in the same way that pressure in a steam engine causes the pistons to move." When the telephone came along, the brain was imagined as a switchboard.

Then, in the 1930s and '40s, thanks largely to the pioneering British mathematician Alan Turing, the seeds for the first computer were sown. "I am building a brain," Turing said of his work. His machine manipulated symbols on a strip of tape, in accordance with a table of rules. It could also compute an important class of mathematical functions. Eventually, our metaphor for the mind became this computerlike machine. Sadly, Turing — whose work was also crucial in the Allies' successful cracking of the Nazi Enigma code — was persecuted for his homosexuality. In 1954, at the age of 41, he died — probably by his own hand — not living to see how fundamentally computers would change the world.

As all these metaphors suggest, humans have been using their brains to think about brains for a very long time. Some of the earliest attempts at neuroscience-type inquiry began more than 6,000 years ago, mostly in the realms of psychoactive drugs and brain disease. The first recorded mention of a word for "brain" appeared in the 1700s BCE, in the *Edwin Smith Papyrus*, a surgical text that provided detailed instructions for treating physical trauma to the body.

In 1664, Thomas Willis coined *neurology*, a term we still use to refer to the branch of medicine that deals with the brain. [13] Willis was an anatomist, but he was also interested in brain function. He discovered several diseases — diabetes, asthma, epilepsy — and he was the first physician to describe hysteria as a brain disease. He was also the first researcher to separate the brain's noodlelike mass into parts, introducing the concept of the cerebral cortex, the outermost layer of gray matter that covers the cerebrum and cerebellum and divides the right and left hemispheres. The cerebral cortex processes the most complex forms of human cognition. It's also the most relevant part of the brain when it comes to understanding how a child's brain intertwines with our nascent Internet.

In the early 1800s, Sir Charles Bell proposed that different sections of the cortex — the brain's powerhouse processor — are specialized for particular functions. Bell called this "cortical localization." Today, scientists are discovering that the connections between the regions are just as important as the regions themselves. In terms of the Internet, this would be like focusing solely on the specific function of each webpage, rather than the way these webpages link together to connect ideas and people.

Bell made his original discovery after dissecting spinal cords. He showed that the cord was divided into motor and sensory regions. In other words, some nerves are for touching, and some are for feeling. Bell also found that the brain's specific regions perform specific tasks. (Over here, you'll find the area responsible for controlling your breathing; over there, the region devoted to voluntary movements.) Bell's 1811 treatise, *Idea of a New Anatomy of the Brain*, has been called the "Magna Carta" of neurology.

In 1849, a German scientist named Hermann von Helmholtz, using the sciatic nerve and calf muscle of a frog's leg, clocked the speed at which a signal is carried along a nerve fiber by recording the fiber's minute electrical impulses. Scientists had believed nerve speed was too fast to measure at all, but von Helmholtz proved it was much slower, clocking it at 24.6 to 38.4 meters per second (55 to 86 miles per hour), about as fast as a 19th-century steam locomotive.

Understanding that nerve fibers transmit electricity at measurable speeds was striking in its own right, but it also meant that the body's language of nerve signals — its Morse code — could be cracked. Scientists soon discovered that different nerve fibers move at different speeds. This explains why rubbing an injured finger makes it hurt less (the cells that tell your brain your finger is being rubbed move faster than the ones telling your brain your finger hurts, thereby overriding the pain signals). Some nerve fibers travel more quickly than others, thus making things happen more rapidly. And a similar thing can happen on the Web by increasing bandwidth and connection speeds. With this analogy, adding bandwidth is like giving data the ability to move more quickly. Just think about the difference between a dial-up connection and high-speed broadband; one set of data is going to get there faster.

Santiago Ramón y Cajal, an accomplished neuroanatomist with expertise in the central nervous systems of many species, came up with a model that became known as the "neuron theory" or "neuron doctrine." **14 15** It said that the brain is made of separate, individual cells (neurons), as opposed to one continuous strand. These cells, Ramón y Cajal found, have specific shapes and functions and are linked by synapses — spaces between neurons where chemical messengers exchange information. This discovery was huge: It meant that neurons could communicate with each other through synapses, even if they weren't touching. It's similar to a wireless connection, the way data is able to jump across networks. You can think of neurons communicating in an analogous way.

In 1909, Korbinian Brodmann, a German scientist, completed the first map of the regions of the brain. Brodmann identified 52 sections, now called the Brodmann Areas, that he found to be different from one another, based on changes in cell type and patterns throughout the cortical surface of the brain. The Brodmann Areas remain the standard classification today. Think of them as zip codes — or, in today's terms, perhaps as a Google map of your brain. Number 17 is the area for sight, numbers 41 and 42 control hearing, numbers 1 to 3 have to do with touch. Understanding specialized functions of the brain allows us to tailor technology, art, and media specifically for the human mind, as we now understand it. Technology is an extension of us — an extension of our minds — and the more we know about the different parts of the brain, the more we're able to develop our technologies to alter how we interact with the world around us.

YOU, ONLY BETTER

The ability to change the brain's structure, which Polish scientist Jerzy Konorski termed "neural plasticity,"[16] was, it was generally believed, only possible during childhood. But research begun in the early part of the 20th century suggested that the brain continually changes throughout life. Though it changes most during the critical years between birth and age 5, the brain is actually evolving all the time.

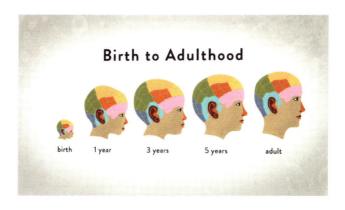

Birth to Adulthood

birth   1 year   3 years   5 years   adult

When considering whether our increasing dependence on technology is good or bad, we can take into consideration the argument scientists make that no matter what, our brains are wired to adapt, and are continuously changing.

So is technology reshaping our neuronal connections? This thinking suggests our brains are wired for that.

Psychologist and brain scientist Karl Lashley was an early pioneer in brain plasticity. In the 1920s, he showed that a monkey could switch hand dominance when grasping for food simply by being rewarded for using its nondominant hand to reach for a food pellet. With training, the monkey would begin to automatically use its nondominant hand to grab food, even without reward. [17] This seemingly simple task was enough to show that repeated changes in the environment could influence the behavior of an adult brain. The repeated actions forced the brain to forge new connections. This finding would hold tremendous potential, suggesting that the brain can rehabilitate and recover from trauma, stroke, or even, to some extent, childhood neglect.

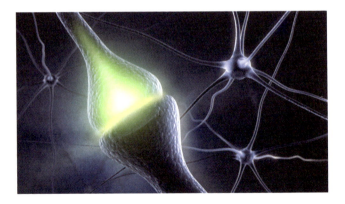

In 1934, Lashley, joined by Ph.D. student Donald Hebb, began experimenting with the effects of environment on brain development in rats. The scientists noted that the brain could develop in different ways, depending on when a rat was first exposed to light. Hebb continued his career in developmental research, studying with the famous neurosurgeon Wilder Penfield. Together, they revealed that the brain of a young child had great potential for adaptation after an injury. In 1949, Hebb proposed the now-famous theory that changes in brain function determine all changes in behavior. Hebb described a theory of neuroplasticity that was summed up by his successors as "fire together, wire together": Neurons that fire in synchrony will strengthen their connections, eventually resulting in new behaviors and learning, whereas neurons that remain out of sync will weaken connections, with information and behaviors gradually lost. In other words, use it or lose it. Hebbian theory has remained one of the strongest explanations for how the human brain learns and how cognition emerges from neural function. It serves as a reminder that everything we do (and don't do) changes the shape of our brain continuously. This includes all the actions we take (like practicing a skill) and all the thoughts and attitudes we have (which are reinforced by repetition).

I SHARE,
THEREFORE I AM

Massachusetts Institute of Technology psychology professor and author Sherry Turkle adapted Descartes' aphorism for the modern age when she said, "I share, therefore I am." She was referring to the growing pervasiveness of sharing one's personal life online. While Turkle is talking about sharing (and, perhaps, oversharing), she and other Internet thinkers don't dispute that there is power in sharing on the Web. Howard Rheingold, who teaches courses on social media and virtual communities at UC Berkeley and Stanford, identified the potential of crowds communicating over the Internet in his 2002 book, *Smart Mobs*, arguing that online groups are able to coordinate in ways never before possible.

Today, when people use technologies like Twitter and text messaging, they can trade and organize information more quickly than ever before. Think of the domino effect of events during the Arab Spring and Occupy Wall Street movements. Or even going back to 2001, remember that within an hour of Philippine President Joseph Estrada's latest corrupt act, a text message was sent that led to protests and his eventual overthrow. When the global brain fires off a message to the local feet, people get out, march, and start either making changes or letting the world know they want them.

There are many other good things that can happen when diverse perspectives come together online. Jeff Howe named this phenomenon "crowdsourcing" in an article he wrote for *Wired* in 2006. One example of using the power of networks is a project that Ken and his students at UC Berkeley are developing, called Opinion Space. It reaches out to large groups for answers to questions by way of a graphical "map" that displays patterns, trends, and insights as they emerge — and then employs the wisdom of crowds to identify and highlight the most insightful ideas. Crowdsourcing is allowing people to do so many new things in new ways — from solving problems and brainstorming to finding financial help. Today artists, filmmakers, and inventors are raising funds for projects by reaching out to their communities and to social media through sites like Kickstarter and Indiegogo. I will never forget when I saw a tweet announcing that Kickstarter had raised more money for its projects that year than the National Endowment for the Arts had given out.

CLOUD
FILMMAKING

YOUR BRAIN
IS THE CLOUD

Last year, at our film studio, The Moxie Institute, we launched what we call Cloud Filmmaking, a new way of crowdsourcing creativity. One of the final scenes in our documentary *Connected* imagines what could happen when everyone who wants to be online is. We call this section of our film the "participatory revolution."

In 1966, psychologist Abraham Maslow famously observed that when the only tool you have is a hammer, you see everything as a nail. To update his maxim for the digital age, if you have a camera in your hands, everything you see tells a story. Right now, more than two billion people are connected online. There are billions of devices with small video cameras that create sendable files. We wondered what would happen if we wrote a script and asked the world to help us create the accompanying film.

We posted the first Cloud Filmmaking script, *A Declaration of Interdependence* (a rewriting of the Declaration of Independence), online on the fourth of July, 2011.

We invited people all over the world to read this in their native language, and then to send us artwork, based on its

A DECLARATION *of* INTERDEPENDENCE

When in the course of human events, it becomes increasingly necessary to recognize the fundamental qualities that connect us,

Then we must reevaluate the truths we hold to be self-evident:

That all humans are created equal and all are connected.

That we share the pursuits of life, liberty, happiness, food, water, shelter, safety, education, justice, and hopes for a better future.

That our collective knowledge, economy, technology, and environment are fundamentally interdependent.

That what will propel us forward as a species is our curiosity, our ability to forgive, our ability to appreciate, our courage, and our desire to connect...

That these things we share will ultimately help us evolve to our fullest common potential.

And whereas we should take our problems seriously, we should never take ourselves too seriously.

Because another thing that connects us...is our ability to laugh...and our attempt to learn from our mistakes.

So that we can learn from the past, understand our place in the world, and use our collective knowledge to create a better future.

We can make the future whatever we want it to be.

So perhaps it's time that we, as a species, who love to laugh, ask questions, and connect.... do something radical and true.

For centuries, we have declared independence.

Perhaps it's now time that we, as humans, declare our interdependence!

words and ideas. Entries poured in from around the globe. We compiled and edited them into a four-minute film, with music by Moby.

A crucial part of Cloud Filmmaking is giving back — again, using the cloud. For each of these films, we offer free, customized versions to any organization working to make the world better, adapting the end of the film to fit that organization's message. For *A Declaration of Interdependence*, we received more than 100 requests from such organizations in the first year. The message was traveling faster, creating new connections around the world.

For *Engage*, the second in our Cloud Film series, we received more than 100 requests in three days.

Brain Power is the third film in the Cloud series. In this film, we wanted to explore how today's technology could help us understand the brain in a new way. So we used that technology to ask people everywhere how they imagine their brains. People started sending us powerful images of their perceptions: some beautiful, some disturbing, some abstract, all exciting. These were like neurons, from every corner of the world, firing ideas and images back at us. They underscored the idea we'd started with: The Internet, the most advanced technological system in the world, is the best framework for understanding the human brain, the most advanced biological system in the world.

Just as Sigmund Freud compared the human brain to a steam engine, and as Alan Turing compared it to his Turing

Machine, today's global community shows us how the brain is like the Internet, the Internet like the brain.

There are others who've envisioned this type of framework. Vannevar Bush, writing in *The Atlantic Monthly* in 1945, imagined the "memex," a "supplement to one's memory." He described it as "a device in which an individual stores all his books, records, and communications, and which is mechanized so that it may be consulted with exceeding speed and flexibility." The paleontologist-philosopher Pierre Teilhard de Chardin's 1955 book, *The Phenomenon of Man*, explores the "noosphere," or the collective human consciousness — the Earth's mental layer formed by the totality of human thought. Peter Russell coined the term "global brain" in his 1982 book of the same name. In it, he expands on Marshall McLuhan's 1960s concept of a global village — a world closely connected by telecommunications. Russell takes McLuhan's global vision further, speculating that new telecommunication technologies will lead to a full awakening of humanity's consciousness.

The two-minute film *Engage* speaks perfectly to this idea. Over the course of six months, we asked people all over the world to send videos of themselves feeling their heartbeat and thinking about what it means. *Engage* is the film that unfolded.

Our experiments in Cloud Filmmaking showed us that our global brain is only getting more interesting, as more people get online, offering more perspectives to further shape its architecture. The Internet is always adding new cells, always adapting, based on the stimulation around it. Both the Internet and the human brain are constantly changing, incredibly complex networks, made up of billions of elements working together and interacting with their environment.

TWO BILLION AND COUNTING

It's strange to think how young the Internet is, considering its enormity and complexity. When I was 17, I fell in love with an Apple computer. My family had an Apple IIe, and I was intoxicated by the idea that a computer and a modem could link me to others. I'll never forget the scratchy, sonic sound when it was connecting — it was as if I could hear the space I was traversing with my computer. This was in the late '80s, before the birth of the Web. But the potential to reach out to people from around the world inspired me, a Russian Jew by heritage, and an Iranian friend of mine to think about how computers could circumnavigate the unidirectional message we were being fed: Both our families came from countries cast as "the enemy."

What if students could communicate directly? We cowrote a proposal called UNITAS, which stood for Uniting Nations in Telecommunications and Software. It posited what might happen if students from adversarial countries could communicate directly through computers to foster a better understanding of our common interests. Because of this proposal, I was invited to the Soviet Union as a student ambassador in 1988. But as soon as I saw how many essentials most people there lacked, I realized that personal computers — and personal connectivity — weren't going to be a global reality for a long time. Two decades later, there are more than two billion people online. In 1996, when I was first running The Webby Awards, there were only around 36 million. Today, it's impossible for me to imagine not being connected via email, not being able to Google any piece of information at any time, from the device in my pocket — even when I'm on an airplane. Our expectations have definitely changed.

In Peter Diamandis' 2012 book, *Abundance: The Future Is Better Than You Think*, he mentions how someone holding a smartphone today has just as much knowledge at their fingertips as a U.S. president did in the 1960s. It's interesting to consider how much more information the youngest generation has than those who've come before it. Think about what the Internet has to offer, compared with radio or early television.

This shift in access to information happened very quickly. In 1992, there was one website. Three years later, there were 623. By 1996, there were 100,000 of them. Today, there are close to a trillion. Shopping online was once considered high-risk. Now, millions of us buy goods from Amazon. It used to be that we felt connected to each other because we had access to email and Listservs. Now, we have thousands of Twitter followers and Facebook friends. With everyone making predictions about where this could lead, Ken and I couldn't help seeing the absurdity in the logical outcome of all this — a world in which we can have seven billion "friends," and the word *friend* itself becomes a verb before becoming completely meaningless. We even made a film, *Facing the Future*, that is a parody of the situation. You can watch it at www.moxieinstitute.org

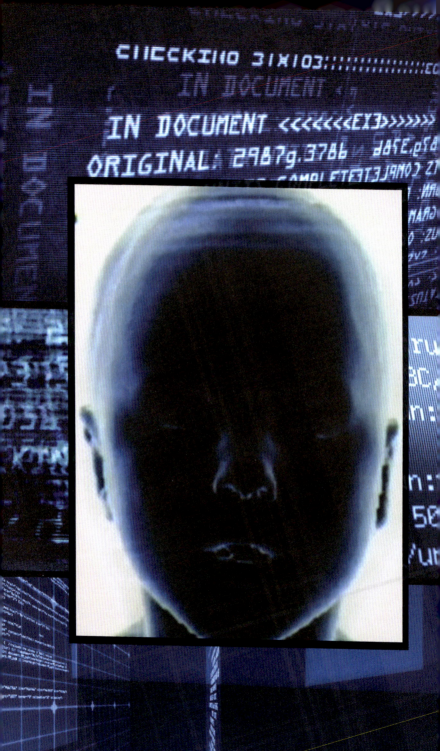

THE INTERNET AS A CHILD

When you start to think about it, the stage of the Internet today is similar to a stage of development of a child's brain. Both a young brain and our young, global Internet brain are in highly creative, experimental, innovative states of rapid development — just waiting to make connections. So, here's a question for the 21st century: How do we help shape both of these young, rapidly growing networks to set a course for a better future? Let's start by diving into this analogy of the brain and the Internet to see what we can learn about both.

While we used to think that what you're born with is what you get, today we know the brain as a network of connections that is constantly growing, expanding, and changing — just like the Internet. With this analogy, a neuron would be like a webpage. Right now, neurons are telling your hands to hold this book, and they're letting your brain know how it feels against your palms. Neurons are directing your eyes to move across the page and are processing the letters into words and meanings. Then they transmit those meanings into memories and new ideas. And they're doing all this via synapses, the connections between them. In the realm of the Internet, webpages act like neurons. Instead of synapses, they have links, which transmit information, make connections, move things around, and make communication happen. So, which would you guess there are more of: neurons in the brain, or webpages on the Internet? Well, a human brain has around a hundred billion neurons. But the Internet has 10 times that — a trillion webpages. **18**

Let's keep exploring this analogy. We could say that a synapse in the brain, the connection point between two neurons, is like a link between webpages. Synapses are like digital wireless connections that allow a neuron to pass a signal to a cell. They are — forgive the pun — sort of like cellphones, an element that lets two cells talk to each other without being hardwired. There are two different types of synapses: electrical, which transmit messages via electrical current; and chemical, which dispatch messages through chemicals known as neurotransmitters. Just like the signal that's emitted from your wireless Internet router to your computer and that enables you to send an email or search the Web, the neurotransmitter initiates an electrical response that excites the postsynaptic neuron. In plain English, the first neuron tells the second neuron something that gets it all fired up.

So, are there more synapses in the brain, or links in the Internet? The Internet has more than a hundred trillion connections. Google's webmaster guideline has recommended a hundred links per page. If each page has a hundred links, that means the Web could have a hundred trillion links. But a brain has three times as many connections — about three hundred trillion. Just think about that. As hard as it is to get your head around how immense the Internet is, try to make it grasp the reality that your brain is actually more so.

These are the moments that fill me with awe, and with humility, about all we still don't know regarding the brain. In our film **Connected**, I wrote that "the brain is the most advanced biological construction on earth." I have thought about the impact of that statement, but it wasn't until we did this computation for **Brain Power** that I fully comprehended the depth of it.

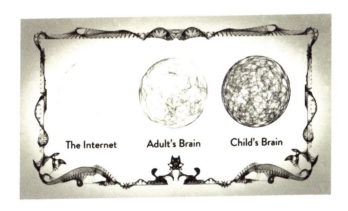

The Internet   Adult's Brain   Child's Brain

Yet, a child's brain has even more connections — a quadrillion of them. That's ten times the number of connections of the entire Internet! Yes, it blew our minds, too. How is that possible? Let's break it down. We're born with a hundred billion neurons — the number we'll have for the rest of our lives. But the connections between those neurons are not there. Then, during the first 2,000 days — about five years — a child's brain grows very rapidly, creating billions upon billions of these connections.

Patricia Kuhl, you'll recall, and her colleagues at I-LABS, study the brain when it's most malleable, from birth to age 5, using MEG imaging. Kuhl and her team are allowing us to see, for the first time, how a child's brain architecture grows as it manipulates the world through play, through listening to language, and through interacting with loving parents and others who are creating the connections and stimulating the processing that leads to thought and creativity. "The child's brain in the first 2,000 days is waiting for this activity," says Kuhl. "And if this activity doesn't occur, the brain doesn't work its magic." It is, she explains, a crucial window.

The current state of the Web parallels this phase of development. There are seven billion people on the planet, but the connections between most of them aren't there yet. As more and more people become plugged in, they can connect, over webpages, in an incalculable number of ways, sharing ideas, information, and innovations.

YOU ARE
YOUR NETWORK

Children's synapses are incredibly prolific, multiplying, multiplying, multiplying. Babies begin connecting with their world from the first moment they're in the world. According to Andrew Meltzoff, who codirects I-LABS with Patricia Kuhl, babies can begin imitating adult actions at less than an hour old. He describes how he poked his tongue out at a brand-new infant, and how she poked her tongue back; how he opened and closed his mouth, and how she did the same things back. "This shows that babies are born social," explains Meltzoff. "They're born connected to us, and this is the root of our common humanity."

I-LABS describes its mission as follows: "The lab has an explicit focus on very young children — and more broadly, understanding what it means to be human through the lens of systems neuroscience, a branch of neuroscience that investigates how small units like cells and molecules interact to form networks and complex systems in order to process information and create our cognitive/conscious experiences."

Meanwhile, Jack Shonkoff, a pediatrician and the director of Harvard University's Center on the Developing Child, is translating the science of early childhood development into more effective strategies to reduce disparities in life outcomes. His center uses advances in neuroscience to document, in policy and practice, the importance of a stable, supportive environment for children's brain development.

Between that early period and puberty, the brain prunes the connections it doesn't need. "We prune a rosebush to strengthen the connections that remain," says Kuhl. "So, as the baby prunes, the remaining connections get stronger." This rapid proliferation of connections is important: These connections form the architecture of the brain, and they are created through every interaction a child has.

Since I've started working on this book, I have found myself making eye contact longer with our 3-year-old, Blooma, trying to teach her bigger words, trying to challenge her and make her laugh more. I watch her playing in the sand, feeling the water, looking at the horizon, negotiating a crab shell, and I think, "She is forging connections between areas of the brain, which has more connections right now than mine."

The types of connections we make throughout life are the focus of a new field of neuroscience, connectomics. And a connectome, somewhat like a genome, is a map, a diagram of all the synapses, or connections, between neurons, in your brain. Picture a subway map — but with billions of different lines. Together, connectomics says, these lines are what make you, well, you. As neuroscientist Sebastian Seung writes in his book on the subject, *Connectome: How the Brain's Wiring Makes Us Who We Are*, "minds differ because connectomes differ. Personality and IQ might also be explained by connectomes. Perhaps even your memories, the most idiosyncratic aspect of your personal identity, could be encoded in your connectome. You are more than your genes. You are your connectome."

Connectomics suggests that nurture and experience are as responsible for who you are as nature is. Seung explains:

> *Genes alone cannot explain how your brain got to be the way it is. As you lay nestled in your mother's womb, you already possessed your genome but not yet the memory of your first kiss. Your memories were acquired during your lifetime, not before. Some of you can play the piano; some can ride a bicycle. These are learned abilities rather than instincts programmed by the genes. Your connectome has been formed by all the things you've seen and done and learned. Which means that you play an active role in its creation. Brain wiring may make us who we are, but we play an important role in wiring up our brains.*

If he's right, you're actively creating your mental map through everything you do, experience, read, even think — plus, you have a lot of agency to change the map, subsequently changing who you are. As Seung theorizes, "Curing mental disorders is ultimately about repairing connectomes. In fact, any kind of personal change — educating yourself, drinking less, saving your marriage — is about changing your connectome."

However, the brain is a fantastically intricate machine, and its fiber connections are only one piece of it. Every part of the brain's architecture is important, and every interaction is important, to how our brains are shaped — especially in those first five years of life. Every time we talk to or engage with a child, we are literally forming a brain, just as, every time we go online, we're forming the global brain. Making connections between the different parts of the brain — so that the neurons can fire together — creates new ideas and promotes creative thinking, just as connecting and sharing and collaborating with others around the world is pruning and cultivating and maturing the Internet brain.

TOXIC
STRESS

WHEN IS TOO MUCH TOO MUCH?

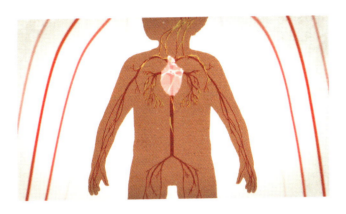

Your baby is crying. You know immediately that something needs to happen: She needs to be fed, or changed, or just held and comforted. If you're like most parents, when your child is in this state, an instinct deep inside tells you to take action. But you should know that their momentary stress is not only bad, but good as well.

Because a child's brain is activated by everything it encounters, it can also be overwhelmed, which causes stress. A little stress is actually good for you, and for your child, because it makes the body and brain go on alert. In fact, it's biologically necessary for survival because it tells us to pay attention. When we are threatened, our bodies activate a variety of physiological responses, including increases in heart rate, blood pressure, and stress hormones, such as cortisol. During these periods, brain signals in the prefrontal cortex become disengaged, which can impair judgment, planning, and decision-making.

If the stress is relieved and a young child receives the support she needs from a caring adult, the stress responses dissipate and the body quickly settles down. When the stress is caused by something beneficial — learning to ride a bike without training wheels, for instance, or going to the doctor — it's called positive stress. When it's caused by something more serious — a hurricane, a divorce, a broken leg — but is well managed, it's called tolerable stress. This type of stress is buffered by caring adults who help the child adapt, which mitigates the potentially damaging effects.

However, when normal, motivating stress turns into so-called "toxic" stress, we need to worry. Jack Shonkoff explains that, in severe situations, such as ongoing abuse and neglect, where there is no caring adult to buffer against the stress, the biological stress response stays activated, and this overloads the body and brain. This has serious lifelong consequences for the child. It results in toxic stress, a bodily stress-response that's permanently set on high alert.

Science now shows that the prolonged activation of stress hormones in early childhood can actually reduce neural connections in important areas of the brain — such as those dedicated to learning and reasoning — while increasing neural connections in the parts of the brain dedicated to fear and aggression. During the crucial early period of development, the brain's circuitry is the most easily influenced, for good or bad, by the child's experiences. If children experience lots of positive, meaningful, stable interaction with the adults in their lives, they build a strong neural foundation for important areas, like those mediating behavior control, language, memory, and motor skills. With repeated use, these circuits become more efficient and connect to other areas of the brain more rapidly. According to "Persistent Fear and Anxiety Can Affect Young Children's Learning and Development," a working paper from the National Scientific Council on the Developing Child (the flagship initiative of Harvard's Center on the Developing Child), research shows that early fears are easier to learn than to unlearn.

Toxic stress experienced early in life — such as that caused by ongoing abuse or neglect, parental substance abuse or mental illness, severe poverty, or recurrent exposure to violence — can have a cumulative toll on a person's physical and mental health. Children who ended up in orphanages right after birth, and who experienced severe neglect, had decreased brain activity when compared with children who were raised in healthy homes.

In a recent Op-Ed for *The New York Times*, Nicholas Kristof wrote about how toxic stress is worse for young children than the threat of guns or speeding cars. According to the website for the Center on the Developing Child, "the more adverse experiences in childhood, the greater the likelihood of developmental delays." Adults who had adverse experiences in childhood are also more likely to have health problems, including alcoholism, depression, heart disease, and diabetes. They are also more likely to struggle in school, to have short tempers, and to have run-ins with the law.

Toxic stress can be avoided if we ensure that the environment in which children grow and develop is nurturing, stable, and engaging. And, as research about neuroplasticity shows, the brain is constantly changing throughout life, so these connections can always change.

Toxic stress isn't restricted to children. But, when it occurs in adulthood, the systems are already in place — and they may be strong or weak, depending on how they developed in childhood, the result of a combination of genetic predisposition and environment. The same stress will affect different people differently. But, with adults, research shows that interventions are less likely to be successful — and, in some cases, they're ineffective.

As Shonkoff observed in *The Encyclopedia on Early Childhood Development*, "When the same children who experienced extreme neglect were placed in responsive foster care families before age 2, their IQs increased more substantially and their brain activity and attachment relationships were more likely to become normal than if they were placed after the age of 2." While there is no magic age for intervention, it is clear that, in most cases, intervening as early as possible is better than waiting.

Toxic stress is a specific and severe human phenomenon, and the Web has no true corollary. But I can't help thinking that metaphorically, there are parallels. While toxic stress is a physiological response our bodies have to ongoing trauma, we can imagine the response our minds might have to ongoing technological overstimulation. If we're only connected to news and media that highlight the worst of humanity, that is going to strengthen certain brain connections more than others. Similarly, if we're always online, and never present with the people right in front of us, we're going to lose some important connections in our brains that have to do with emotion, trust, and empathy. So, while the outcomes are not the same as toxic stress, the way we use technology does affect us.

We could also imagine a metaphoric version of toxic stress for our global brain that is the Internet. Two bills that were on the verge of passing into law in early 2012 (PIPA, the Protect IP Act, and SOPA, the Stop Online Piracy Act — both sound like they'd be good things) would severely limit the free exchange of ideas over the Internet in a very Big

Brother way, ultimately reshaping the Web. If passed, only certain kinds of connections would be permitted, and the government and copyright-owning corporations would have the major say in what those kinds of connections would be. Other information and connections would be limited, or severely neglected. When PIPA threatened to become law, the Internet did what it does best, organizing massive protests. On January 18, 2012, more than 7,000 websites, including Wikipedia, went dark. Google stayed up, but used its home page to promote the protest. As a result, the Senate agreed to postpone a vote until the bill's issues were addressed.

"The idea that all Internet traffic should be treated equally is known as network neutrality," wrote Mathew Honan in *Macworld*. In other words, no matter who uploads or downloads data, or what kind of data is involved, networks should treat all of those packets in the same manner. To do otherwise, advocates argue, would amount to data discrimination. So in our metaphor, net neutrality, which promotes the idea of an "open Internet," when applied to the brain means that all the different parts of the brain can interact and engage in a healthy way.

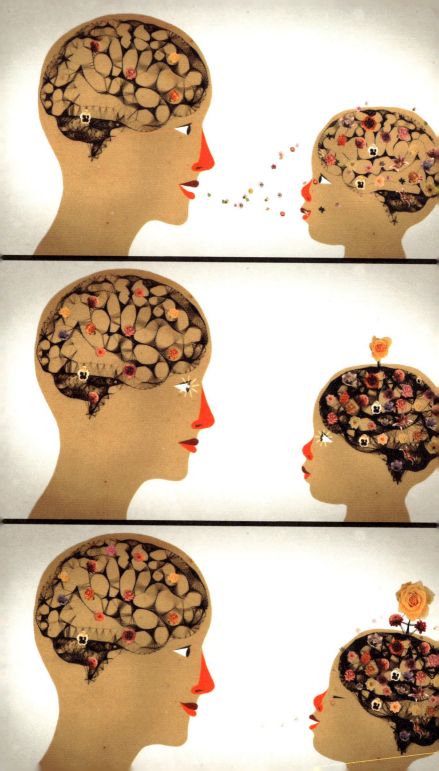

MINDFUL TECHNOLOGY

The South Africans have a beautiful philosophy called Ubuntu, which translates as "I am what I am because of who we all are." This is a perfect way to think about the way a brain develops, influenced by its surrounding people and experiences. It's also how we should think about the way the Internet is developing, and about the way our choices in how we use technology are shaping this global brain. For both the brain and the Internet, the networks are always binding us in new ways and changing our understanding of who we are and how we perceive the world. If we say that the Internet is in the same critical stage of early development as a child, making as many connections as possible, then we need to be mindful of how we're building its foundation.

In his latest book, *Net Smart*, Howard Rheingold outlines five skills for mindfully connecting online: attention (focusing on what's relevant); participation (being a good Internet contributor); collaboration (working with a diverse online community to develop new ideas); critical consumption of information (or, as he calls it, "crap detection"); and network smarts (learning about and building networks). As Rheingold argues, "There is a bigger social issue at work in digital literacy, one that goes beyond personal empowerment. If we combine our efforts wisely, it could produce a more thoughtful society: countless small acts like publishing a Web page or sharing a link could add up to a public good that enriches everybody."

We need to think about how we're shaping the Internet on a global scale. What connectomics are we creating in the global network? Just as every interaction creates new connections in a child's brain, every email, tweet, search, or post is creating and strengthening connections in our global brain, changing the shape of the Internet that we, billions of people, are developing together. How can we mindfully connect people who want to connect, and who want to participate in this global network of knowledge and ideas?

One person working to find the answer is Tim Berners-Lee, the creator of the World Wide Web. Berners-Lee is also the founding director of the World Wide Web Foundation, which aims to increase Web access and usefulness globally. The foundation's three initiatives include Web Rights (which promotes the availability of affordable, uncensored, unmonitored access to the Web), Web for Civic Engagement (which strives to increase civic participation, accountability, and openness), and Web for Economic Development (which encourages the growth of an economic "Web ecosystem" and uses technology to help combat poverty). This is what proceeding mindfully looks like: working together to build a Web that serves vital, neglected needs, that links all who want to be linked, and that fosters creative interaction to make a better world.

Vint Cerf, considered one of the co-creators of the Internet's architecture, wrote in a 2012 *New York Times*

Op-Ed that "The Internet stands at a crossroads. Built from the bottom up, powered by the people, it has become a powerful economic engine and a positive social force. But its success has generated a worrying backlash. Around the world, repressive regimes are putting in place or proposing measures that restrict free expression and affect fundamental rights. Like almost every major infrastructure, the Internet can be abused and its users harmed. We must, however, take great care that the cure for these ills does not do more harm than good. The benefits of the open and accessible Internet are nearly incalculable and their loss would wreak significant social and economic damage."

THE WORLD, ONLINE AND OFF

As of December 31, 2011, 2.4 billion people were connected to the Internet. But think about that in terms of where we could be. It means that 4.6 billion people aren't connected to the Internet, that some 66 percent of the world's population doesn't have access to this global hub of information and are therefore unable to contribute fresh perspectives and new knowledge. If we want to take the Internet and the concept of a global brain to their fullest potential, we have to give as many people as possible the ability to access it. To have the most creative, insightful thoughts and solutions to today's biggest challenges requires the most people possible online — the most ideas, the most capabilities.

Just as it's key that all the different parts of a child's brain are connected in order to set the stage for the most insightful and creative thoughts, it's key that all the different parts of the world are connected, laying the foundation for worldwide empathy, innovation, and human expression. In a child's brain, Kuhl talks about how you need as many different parts connected as possible to maximize "insights and creative ideas." This is also true for the global brain.

Perhaps, like human brains, the global brain will proliferate to connect everything, then will eventually "prune" the connections that don't have long-term benefits and strengthen the connections that are making our global brain stronger — with more sharing of perspectives, more developing of creative ideas for a sustainable future. For both the Internet and a child's brain, the connections we pay the most attention to will be strengthened, while the ones we use less will be pruned.

As we move to a point where everyone can be online all the time, it is just as important to the development of our global brain to know when to disconnect. Just as it's not good for a child's brain to be constantly overstimulated, that's not good for adults' brains, either. Have you noticed that, when you ask people how they are, the answer is almost always "busy"? They're not so much complaining as bragging — or simply reporting, almost like a status update on Facebook. These days, it's a status symbol to be busy. As therapist Ira Israel wrote in a recent Huffington Post blog, it's now considered good to be "crazy busy," which means simultaneously having back-to-back meetings and taking a pregnant friend to the hospital — all while running a conference call. We have become a hive of bees, buzzing with busyness. When was the last time you heard someone say "I'm not really up to anything," or "I'm feeling well balanced in all the things in my life"?

But while it's a good thing to have a full and productive life, too much "busyness" is not a good thing, especially when it prevents real, deep thinking. Maybe what we need is a "busy signal," a reminder to hang up, unplug, and just be for a bit. Or, as Andrew Keen, the author of *Digital Vertigo*, has declared, "Sometimes it's good to just shut up."

Just as we're learning how important it is not to subject the developing brain to toxic stress, we also need to learn how important it is to watch out for technological overstimulation as we develop the Web. This is what technology Shabbats are all about. Just as we educate ourselves about what constitutes too much sugar or too much alcohol, we all need to learn what makes for good connecting — what's healthy as opposed to what's addictive and unhealthy — and, therefore, when to go offline. Strengthening the foundation of the Internet is both a local and a global job. This means that we all need to consider our own personal relationship with technology, how we engage with it or disengage from it. In his new book, *Program or Be Programmed*, Douglas Rushkoff offers his "ten commands" for mastering the role of technology in our lives, rather than letting ourselves be mastered by it. "Choose the former and you gain access to the control panel of civilization," he writes. "Choose the latter, and it could be the last real choice you get to make."

Howard Rheingold suggests something quite simple but revelatory to his students at the beginning of his classes: he asks them to shut their laptops, close their eyes, and spend 60 seconds noticing how ideas arise and float away without conscious intent. Rheingold calls it "paying attention to what you're paying attention to."

Attention is the mind's most valuable resource.

In our whirling technological world, attention is becoming a resource as precious as oil or gold, and we should respect and value it. In her 2008 book, *Distracted: The Erosion of Attention and the Coming of the Dark Age*, Maggie Jackson calls "deep, sustained, perceptive attention" the "building block of intimacy, wisdom, and cultural progress." Jordan Grafman, a member of the Dana Alliance for Brain Initiatives and chief of cognitive neuroscience at NIH's National Institute of Neurological Disorders and Stroke, argues that if technology is used without restraint or judgment, it will shape the brain in a negative way. "A lot of what is appealing about all these types of instant communications is that they are fast," he has said. "Fast is not equated with deliberation. So I think they can produce a tendency toward shallow thinking. It's not going to turn off the brain to thinking deeply and thoughtfully about things, but it is going to make that a little bit more difficult to do."

We are — actively, daily — affecting the connections in our brains as we plug into our smartphones and tablets and laptops. And since it's having such an impact on our minds, we need to do it mindfully, and sometimes not do it at all. Every person you follow on Twitter, or are friends with on Facebook, is influencing your thoughts and ideas — and sometimes they even show up in your dreams. On Twitter, people I follow almost become my stream of consciousness; they are streams of thought in my head. We need to be mindful of who and what we let into our brains. The health of our own brains and our global brain depends on it.

GLOBAL
MIND–MELD

Ultimately, we're social creatures who need authentic emotional connections. We can't survive without them. If the Web is to be meaningful and useful in the future, it, too, must foster meaningful connections. It won't be long before the Internet is all grown up. Let's make it a useful, inspiring, likable entity that we want to spend time with — and that just might make us better people.

In writing this book, and in making the film *Brain Power*, my goal has been to help people think mindfully about what our interactions with our children can teach us with regard to how we are growing our global Internet brain — and vice versa.

Policy makers want to make communities safe for children, improve their learning, and reduce risks to their health. A child raised in such an environment will experience less stress and have more opportunities to make the positive connections that will lead to better brain connections, to more empathy and creativity. This, in turn, will lead to more opportunities for learning, better health, and safer communities. It's a beautiful cycle.

Let's take these same ideas and apply them to the developing global brain. The result? A more connected Web and world. Just as in the brain, connections lead to capability. The more connections we have to more parts of the world, the more capability, insights, and creativity we generate together. So, how do we nurture these two growing and interconnected networks to set the course for a better future? As Rheingold says, by paying attention to what we are paying attention to. Every time we interact with either a child's brain or the global brain, we are shaping its future, and, therefore, the future of the world. Every interaction counts. The process is constant, and constantly changing. With every interaction, we're forming new connections: with our children, with our Web communities, with our own neurons. We're constantly reshaping our connectomes.

By reading this book and watching our film, you've reshaped your brain, too. And we hope you'll go out and talk to or play with a child, or talk or tweet or blog about the ideas in this book. These actions can reshape the connections in a child's brain and thus the global brain, as well as the brains of those who connect with either of those. They will go on to connect with others, who will go on to connect with even more people. All these new synapses are coming from this one junction. Let the ripple effects begin.

Just imagine the brain power.

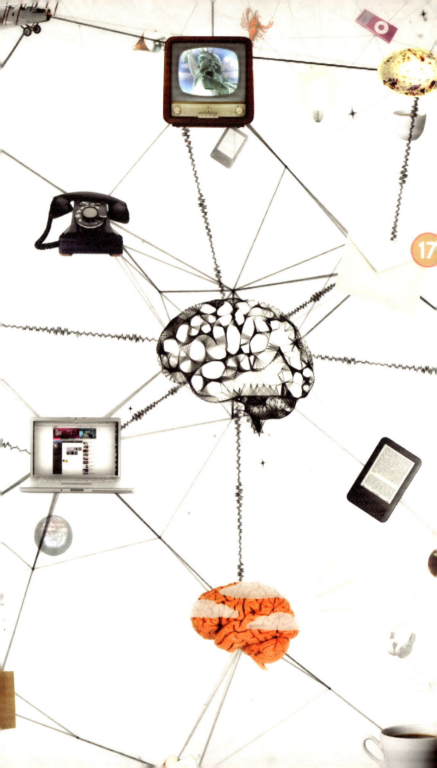

ENDNOTES

1   Jebb, Richard C., *Sophocles: Antigone*, Part 3 (London: Bristol Classical, 2004).

2   whatmakesthemclicknet/2009/11/07/100-things-you shouldknow-about-people-8-dopamine-makes-us-addicted-to-seekinginformation/.

3   Chocano, Carina, "Pinterest, Tumblr and the Trouble With 'Curation,'" *The New York Times Magazine*, July 22, 2012, nytimes.com/2012/07/22/magazine/pinterest-tumblr-and-thetrouble-with-curation html?pagewanted=all.

4   Richtel, Matt, "Attached to Technology and Paying a Price," *The New York Times*, June 6, 2010, nytimes.com/2010/06/07/technology/07brain.html?src=me&ref=general.

5   "First Steps to Digital Detox," Room for Debate, *The New York Times*, June 7, 2010, roomfordebate.blogs.nytimes.com/2010/06/07/first-steps-to-digital-detox/.

6   Kedrosky, Paul, "The Large Information Collider, BDTs, and Gravity Holidays on Tuesdays," edge.org/q2010/q10_7.html.

7   Shlain, Leonard, *The Alphabet versus the Goddess* (Penguin Books, 1999).

8   Baroncelli, L., C. Braschi, M. Spolidoro, T. Begenisic, A. Sale, and L. Maffei, "Nurturing Brain Plasticity: Impact of Environmental Enrichment," Cell Death & Differentiation 17 (7): 1092–1103. doi:10.1038/cdd.2009.193, 2010.

9   Heschel, Abraham Joshua, *The Sabbath* (Farrar Straus Giroux, 2005).

10   Durant, Will, and Ariel Durant, "Books: The Great Gadfly" (review of The Age of Voltaire, by Will and Ariel Durant). *Time magazine*, October 8, 1965.

11   Pearn, John., "A Curious Experiment: The Paradigm Switch from Observation and Speculation to Experimentation, in the Understanding of Neuromuscular Function and Disease," Neuromuscular Disorders 12 (6) (August): 600–607. doi:10.1016/S0960-8966(01)00310-8. 2002.

12   Descartes, René, and Thomas Steele Hall, *Treatise of Man*, 1st ed. (Prometheus Books, 2003).

13   faculty.washington.edu/chudler/hist.html.

14   "Neuron Theory." Wikipedia, Wikimedia Foundation, June 5, 2012. Accessed October 2, 2012. en.wikipedia.org/wiki/Neuron_theory.

15   Golgi, Camillo, "The Neuron Doctrine: Theory and Facts," Nobel lecture, December 11, 1906. nobelprize.org/nobel_prizes/medicine/laureates/1906/golgi-lecture.pdf.

16   Konorski, Jerzy, *Conditioned Reflexes and Neuron Organization*, vol. xiv (New York: Cambridge University Press, 1948).

17   Lashley, Karl S. "Modifiability of the Preferential Use of the Hands in the Rhesus Monkey." *Journal of Animal Behavior 7* (n.d.): 178-86. Print.

18   Sutter, John. "How Many Pages Are on the Internet?" CNN.com. September 12, 2011. cnn.com/2011/TECH/web/09/12/web.index/index.html.

ABOUT THE AUTHOR

Honored by *Newsweek* as one of the "women shaping the 21st century," Tiffany Shlain is a filmmaker, the founder of The Webby Awards, and a co-founder of the International Academy of Digital Arts and Sciences. Tiffany's films and other work have received 50 awards and distinctions. In 2011, she delivered the keynote commencement address at UC Berkeley. Her last four films premiered at Sundance, including her 2011 documentary, *Connected: An Autoblogography about Love, Death & Technology*, which *The New York Times* hailed as "examining everything from the Big Bang to Twitter." *Connected* was selected by the State Department as one of the films to represent America around the world for the 2012 American Film Showcase. *Brain Power* was also selected for 2013-2014 American Film Showcase. Tiffany is currently working on the film series *Let It Ripple: Mobile Films for Global Change*, 16 shorts that look at what connects us as humans. The project employs Cloud Filmmaking, an approach she developed that enlists the participation of people all over the Web. In 2012, she published "The Cloud Filmmaking Manifesto" at the Tribeca

Film Festival, where she received a Disruptive Innovation Award for her film work. Tiffany is a Henry Crown Fellow at the Aspen Institute and a member of the advisory boards for the MIT-IBM Network Science Research Center and for the Institute for the Future. She was invited to advise former Secretary of State Hillary Clinton about the Internet and technology. She lives in the Bay Area with her husband and two children. To check out the other films mentioned in this book, please visit tiffanyshlain.com. Follow on twitter @tiffanyshlain

About**TED**

TED is a nonprofit devoted to "Ideas Worth Spreading."
It started out, in 1984, as a conference bringing together
people from three worlds: Technology, Entertainment,
Design. Since then its scope has become ever broader.
Along with two annual conferences -- the TED Conference
in Long Beach and Palm Springs, Calif., each spring, and
the TEDGlobal conference in Edinburgh, Scotland, each
summer -- TED includes the award-winning TEDTalks video
site, the Open Translation Project and Open TV Project, the
inspiring TED Fellows and TEDx programs, and the annual
TED Prize.

The annual TED Conferences, in Long Beach/Palm Springs
and Edinburgh, bring together the world's most fascinating
thinkers and doers, who are challenged to give the talk of
their lives (in 18 minutes).

On TED.com, we make the best talks and performances
from TED and partners available to the world, for free.
More than 1,000 TEDTalks are available online, with more
added each week. All are subtitled in English; many are
subtitled in various other languages. These videos are
released under a Creative Commons license, so they can
be freely shared and reposted.

TED Conferences, LLC
250 Hudson Street
New York, NY 10013
TED.com

Published simultaneously in the United States and
wherever access to Amazon, the iBookstore, and Barnes &
Noble is available. Also available in TED Books app edition.
First edition. First published November 2012. ISBN-978-1-
937382-21-6.

TED is a registered trademark, and the TED colophon is a
trademark of TED Conferences, LLC.

CPSIA information can be obtained
at www.ICGtesting.com
Printed in the USA
LVIC04n2152100317
526814LV00001B/4

* 9 7 8 0 9 9 8 1 6 8 6 4 7 *